School Library
Blair Ridge P.S.
100 Blackfriar Ave.
Brooklin, ON L1M 0E8

631.53
JOH

GROWING NEW PLANTS

by Terry Johnson

CHERRY LAKE PUBLISHING * ANN ARBOR, MICHIGAN

Published in the United States of America by Cherry Lake Publishing
Ann Arbor, Michigan
www.cherrylakepublishing.com

Content Adviser: Paul Young, MA, Botany

Reading Consultant: Cecilia Minden-Cupp, PhD, Literacy Specialist and Author

Photo Credits: Cover and page 4, ©Katrina Leigh, used under license from Shutterstock, Inc.; page 6, ©iStockphoto.com/mmmphoto; cover and page 8, ©Khorkova Olga, used under license from Shutterstock, Inc.; page 10, ©Milos Luzanin, used under license from Shutterstock, Inc.; cover and page 12, ©Polina Lobanova, used under license from Shutterstock, Inc.; cover and page 14, ©Garden World Images Ltd/Alamy; page 16, ©Fotocrisis, used under license from Shutterstock, Inc.; page 18, ©N Joy Neish, used under license from Shutterstock, Inc.; page 20, ©Volodymyr Pylypchuk, used under license from Shutterstock, Inc.

Copyright ©2009 by Cherry Lake Publishing
All rights reserved. No part of this book may be reproduced or utilized in any form
or by any means without written permission from the publisher.

LIBRARY OF CONGRESS CATALOGING-IN-PUBLICATION DATA
Johnson, Terry, 1964–
  Growing new plants / by Terry Johnson.
    p. cm.—(21st century junior library)
  Includes index.
  ISBN-13: 978-1-60279-279-1
  ISBN-10: 1-60279-279-8
  1. Plants—Juvenile literature. 2. Plant propagation—Juvenile
literature. 3. Gardening—Juvenile literature. I. Title. II. Series.
  SB406.7.J64 2008
  631.5′3—dc22   2008014564

---

*Cherry Lake Publishing would like to acknowledge the work of
The Partnership for 21st Century Skills.
Please visit www.21stcenturyskills.org for more information.*

CONTENTS

5   **Same or Different?**

11   **Ways to Grow**

17   **Plants on the Go**

22   Glossary

23   Find Out More

24   Index

24   About the Author

Plants need room for their roots to grow.

Same or Different?

Plants have many things in common. They need the same things to grow healthy and strong. They need sunlight, air, and water. They need room for their roots to grow.

Roses are just one kind of perennial plant.

Roots take in **minerals** and water for the plant. The plant stem carries the minerals and water to all the parts of the plant. Leaves use sunlight, air, and water to make food for the plant.

Plants are also different. Some plants, such as trees, grass, and roses, seem to die during the winter. But they come back year after year. These plants are called **perennials**. Other plants grow, flower, make seeds, and die in one year. These plants are called **annuals**.

Marigolds are annuals. They grow, flower, make seeds, and die in one year.

Plants grow new plants in different ways, too. Let's take a look at the ways new plants can grow.

Create! Cut out pictures of five plants from old magazines. Tape them to a big piece of cardboard. Look in books or have an adult help you search online. Find out if the plants are annuals or perennials. Label them. Proudly display your plant knowledge!

The seeds of many plants grow inside fruits. Do you see the seeds inside this apple?

10

Ways to Grow

Plants can make new plants like themselves. This is called **reproduction**. How do plants reproduce? They can do this in several different ways.

Some plants reproduce by making seeds. Flowers contain the parts of plants that make seeds. Seeds can be planted in soil. Then they will grow into new plants if they get enough water and sunlight.

Some plants grow from bulbs.

Some plants don't have flowers. They produce tiny **spores** instead of seeds. New plants can grow from the spores. A fern is one kind of plant that makes spores to reproduce.

Some plants grow from **bulbs**. Tulips and daffodils are two kinds of plants that grow from bulbs. A bulb looks a little bit like an onion. It contains a tiny baby plant. The bulb also contains the food that will feed the tiny plant as it grows.

Sometimes cuttings are placed in water until roots begin to grow. Then they are planted in soil.

Some plants can grow from **cuttings**. A cutting is a small piece cut from the stem or a leaf of a plant. It will grow roots and become a new plant if it is planted in soil.

Sometimes a cutting from a tree can be joined to the stem of an older tree. Then the two plants become one plant. This is called **grafting**.

Make a Guess!

Think of three different kinds of plants. Make a guess about whether each plant grows from seeds, spores, or bulbs. Look in books or go online with an adult to find out if your guesses were right.

Dandelion seeds are made to blow easily in the wind

Plants on the Go

What do wind, water, humans, and animals have in common? They all move parts of plants from one place to another to make new plants!

Wind and water can move plant parts from one yard or even one farm to another. A strong wind can carry seeds for many miles.

People who plant gardens often move plants from one place to another.

Sometimes people help move plants. They can separate a clump of roots in half. Then they replant each half so they have two plants. People also take cuttings of plants. Then new plants can grow from the stem or leaf cutting. People also plant seeds to grow new plants.

Birds and insects move **pollen** from one flower to another. The dust-sized pollen grains help the plant make more seeds.

Can you see the yellow pollen grains on the bee's legs?

Take a walk, and look at the plant world around you. The plants you see came from seeds, bulbs, stems, leaves, roots, and spores.

Why not plant a garden? Then you can grow some new plants of your own!

Find a flower. Tulips and poppies work well. Ask an adult before you pick one! Gently pull off the petals. Can you see the powdery pollen grains? Use a magnifying glass if your flower is small.

GLOSSARY

annuals (AN-you-uhlz) plants that grow, flower, make seeds, and die in one year

bulbs (BUHLBZ) a small hard lump of plant matter that contains a tiny baby plant and the food it needs to grow

cuttings (KUT-ingz) small pieces of a plant leaf or stem that can be used to grow a new plant

grafting (GRAF-ting) joining a piece of one plant onto another so they grow into one plant

minerals (MIN-ur-uhlz) substances found in nature that aren't alive but are needed in small amounts by many living things

perennials (purr-EN-ee-uhlz) plants that live and flower year after year

pollen (POL-uhn) dust-sized grains that help make seeds in flowers

reproduction (ree-pruh-DUK-shuhn) the act of producing new plants or animals like their parents

spores (SPORZ) tiny parts of some plants that can grow into a new plant

FIND OUT MORE

BOOKS

Robbins, Ken. *Seeds*. New York: Atheneum Books for Young Readers, 2005.

Weiss, Ellen. *From Bulb to Daffodil*. Danbury, CT: Children's Press, 2007.

WEB SITES

My First Garden—Show Me the Basics
www.urbanext.uiuc.edu/ firstgarden/basics/index.html
Basic information on gardening for kids

Zoom—Science Rocks: Sock Seeds
pbskids.org/zoom/activities/sci/ sockseeds.html
Try a fun activity and grow plants using socks

INDEX

A
air, 5, 7
animals, 17, 19
annuals, 7, 9

B
birds, 19
bulbs, 13, 15, 21

C
cuttings, 15, 19

D
daffodils, 13

F
ferns, 13
flowers, 7, 11, 13, 21
food, 7, 13

G
gardens, 21
grafting, 15
grass, 7

H
humans, 17, 19

I
insects, 19

L
leaves, 15, 19, 21

M
minerals, 7

P
perennials, 7, 9
petals, 21
pollen, 19, 21
poppies, 21

R
reproduction, 11, 13
roots, 5, 7, 15, 19, 21
roses, 7

S
seeds, 7, 11, 15, 17, 19, 21
soil, 11, 15
spores, 13, 15, 21
stems, 7, 15, 19, 21
sunlight, 5, 7, 11

T
trees, 7, 15
tulips, 13, 21

W
water, 5, 7, 11, 17
wind, 17

ABOUT THE AUTHOR

Terry Johnson is a Master Gardener. She enjoys yoga, reading, and antiquing. She lives with her husband and black Labrador retriever near Memphis, Tennessee.